Hurricane

Catherine Chambers

Heinemann Library
Chicago, Illinois

Customer Service 888-454-2279

Visit our website at www.heinemannlibrary.com

Designed by Visual Image
Illustrations by Paul Bale
Originated by Ambassador Litho
Printed and bound in South China

06 05 04 03 02
10 9 8 7 6 5 4 3 2 1

Library of Congress Cataloging-in-Publication Data
Chambers, Catherine, 1954-
  Hurricane / Catherine Chambers.
    p. cm. -- (Wild weather)
Includes bibliographical references index.
Summary: Describes how hurricanes are formed, how they are measured, the
harmful and beneficial impact of these storms, and their impact on
humans, animals, and plants.
  ISBN 1-58810-651-9(HC), 1-4034-0114-4 (Pbk)
  1. Hurricanes--Juvenile literature. 2. Hurricanes--Physiological
effect--Juvenile literature. [1. Hurricanes.] I. Title. II. Series.
  QC944.2 .C478 2002
  551.55'2--dc21
                        2002000821

Acknowledgments

The author and publishers are grateful to the following for permission to reproduce copyright material: pp. 4, 7, 9, 13
Associated Press; p. 17 Colorific; pp. 5, 11, 16, 20, 23, 24, 27, 29 Corbis; p. 21 PA Photos; p. 28 Panos; pp. 14, 25
Photodisc; pp. 15, 19, 22, 26 Rex Features; p. 12 Robert Harding Picture Library; p. 10 Science Photo Library; p. 18 Stone.

Cover photograph reproduced with permission of Pictor.

The Publishers would like to thank the Met Office for their assistance in the preparation of this book.

Every effort has been made to contact copyright holders of any material reproduced in this book. Any omissions will be
rectified in subsequent printings if notice is given to the publisher.

Some words are shown in bold, **like this.** You
can find out what they mean by looking
in the glossary.

Contents

What Is a Hurricane?

A hurricane is a huge storm that builds up over an **ocean.** If a hurricane reaches the land, it can bring strong winds and heavy rain.

Hurricane winds can smash windows and blow roofs off houses. They can also snap trees and flatten **crops.** A hurricane's heavy rain can cause **floods.**

Where Do Hurricanes Happen?

Most hurricanes happen in an area called the **Tropics.** The Tropics are hot because the heat of the Sun is stronger in this region. Hurricanes are called "typhoons" in some parts of the world.

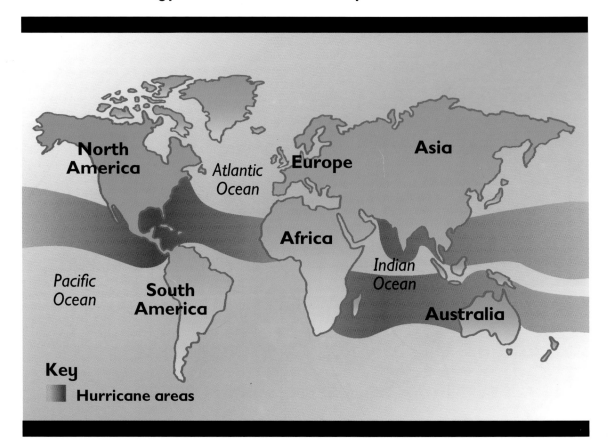

North America

Atlantic Ocean

Europe

Asia

Africa

Indian Ocean

Pacific Ocean

South America

Australia

Key

Hurricane areas

The country of Honduras is in Central America. Very strong hurricanes often happen here because it is close to warm **oceans.** Warm oceans help hurricanes to form.

How Do Hurricanes Form?

Masses of air are always moving. A warm mass of air usually rises. This makes a low **pressure** area. A cold mass sinks. This makes high pressure. Winds blow from high pressure to low pressure.

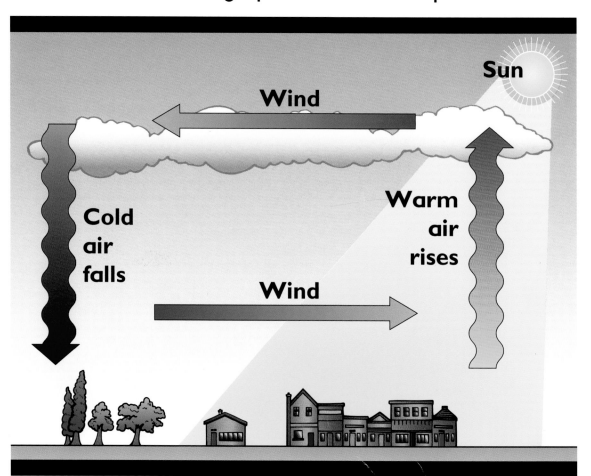

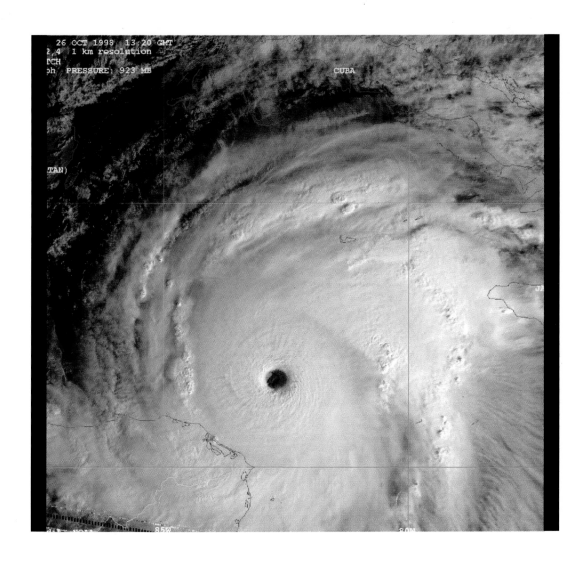

Hurricanes happen when air rises quickly over warm **oceans** in the **Tropics.** This makes a low pressure area. Strong winds rush in from high pressure areas. They form a **spiral** of wind and clouds around an area of calm air.

What Do Hurricanes Do?

Hurricanes are very powerful storms. Their winds blow the **ocean** into huge waves. Heavy rain falls from the huge, dark clouds that make up the hurricane.

Many hurricanes blow over tropical **coasts.**
As a hurricane moves over the land, it gets less
powerful. After a while, it will die out. But it can
do a lot of **damage** before it does.

11

What Are Hurricanes Like?

Sometimes hurricanes pass through towns and cities. Their strong winds may uproot trees and blow roofs off buildings. It can get so windy that it is not safe to go outside.

Hurricanes also bring heavy rain. They pick up water from the **ocean** and drop the water onto the land. This often causes **floods.**

Harmful Hurricanes

A hurricane **storm surge** has hit this **coast.**
A storm surge is a huge wave made by strong
winds. This wave can then **flood** the land,
carrying boats with it.

Hurricanes can **damage** roads, railroads, bridges, and airports. This makes it difficult for rescuers to reach people after a hurricane.

Hurricane in Jamaica

This is the island of Jamaica. It is a country that lies in the Caribbean Sea. Many people visit Jamaica because of its beautiful beaches. But Jamaica is in a hurricane area.

Hurricane Gilbert swept across Jamaica in 1988.
It **damaged** many buildings, roads, and bridges.
Some were completely destroyed. The hurricane
also damaged **crops** across the country.

17

Preparing for a Hurricane

Weather **forecasters** look at photos taken by **satellites** in space. They can use these photos to track hurricanes as they form and move toward land.

People can watch
television or listen to
the radio to hear
hurricane warnings.
On the **coast,** special
flags are sometimes
hung. These warn
people that a hurricane
is on the way.

Coping with Hurricanes

People often nail strong wooden boards in front of doors and windows before a hurricane strikes. The boards keep the hurricane's winds from smashing the windows.

Many people go to community centers or special **shelters** before the hurricane hits. They are given food and a place to sleep. They stay until the hurricane is over and it is safe to go home.

Hurricane Andrew

Hurricane Andrew arrived in the United States in 1992. The hurricane hit parts of Florida and Louisiana. People got ready for the storm as well as they could, but it was too strong.

Hurricane Andrew **damaged** tall office buildings and an Air Force base. It caused more damage than any other hurricane in U.S. history.

Nature and Hurricanes

Hurricane winds and rain can destroy **crops.** In poor countries like Bangladesh, these crops are very important. If they are destroyed, people will not have enough to eat.

Palm trees grow in the **Tropics.** These trees can bend and sway without breaking. They can stand up to the strong winds of a hurricane.

To the Rescue!

People can get trapped inside their houses if a hurricane **damages** them. Rescuers help people buried under fallen houses and trailers. Trailers get damaged easily by the strong winds.

Heavy rain can **flood** drains and water supplies. This makes the water dirty. Dirty drinking water can cause **disease.** So it is important to bring in fresh water after a hurricane.

Adapting to Hurricanes

Belize is a country in Central America. In 1961 a hurricane destroyed its **capital** city, which was on the **coast.** A new city was then built inland, where it would be safe from hurricanes.

Parts of Australia sometimes get hurricanes.
People in those areas now have to build
stronger buildings that will not be destroyed
in a hurricane.

Fact File

◆ The worst known hurricane disaster happened in the country of Bangladesh in 1970. About 500,000 people died.

◆ There is a small area of calm air in the center of a hurricane. This place is called the "eye" of the storm.

◆ Every hurricane is given a name such as Mitch or Pauline. Scientists assign the names in alphabetical order. If one hurricane has a boy's name, the next will have a girl's name, and so on. They are not named after real people.

Glossary

capital most important city of a country, where the government meets

coast area where the land meets the sea

crop plant that is grown for food

damage to harm something; or, the harm that is caused

disease serious illness

flood overflow of water onto a place that is usually dry

forecaster someone who collects information about the weather in order to predict what kind of weather we will get

mass large amount of something like air that does not have a definite shape

ocean one of the four large bodies of salt water that cover Earth

pressure pushing force

satellite spacecraft that travels around Earth

spiral shape that is made by circling around a center, getting bigger and bigger

storm surge huge sea wave pushed to the shore by hurricane winds

Tropics very warm areas that lie near the Equator

More Books to Read

Ashwell, Miranda, and Andy Owen. *Wind*. Chicago: Heinemann Library, 1999.

Gentle, Victor, and Janet Perry. *Hurricanes*. Milwaukee: Gareth Stevens, 2001.

Morgan, Sally. *Read about Hurricanes*. Brookfield, Conn.: Millbrook Press, 2000.

Index

DATE			